MY FIRST DEPLOYMENT

CHRONICLES OF A MILITARY JOURNEY

RONALD BETHUNE

My First Deployment:

Chronicles of a Military Journey

Ronald Bethune

2024 © by Ronald Bethune

All rights reserved. Published 2024.

No part of this publication may be reproduced or transmitted in any form or by any means, electronic or mechanical, including photocopying, recording, or any information storage and retrieval system now known or yet to be invented, without written permission from the publisher, except in cases of brief quotations in critical articles and reviews.

BIBLE SCRIPTURES

Scripture quotations marked (NLT) are taken from the Holy Bible, New Living Translation (NLT), Copyright © 1996, 2004, 2015 by Tyndale House Foundation. Used by permission of Tyndale House Publishers, a Division of Tyndale House Ministries, Carol Stream, Illinois 60188. All rights reserved.

Printed in the United States of America

Spirit Media and our logos are trademarks of Spirit Media

SpiritMedia.US
www.spiritmedia.us
8045 Arco Corporate Dr STE 130
Raleigh, NC 27617
1 (888) 800-3744

Books › Religion & Spirituality › Christian Books & Bibles › Christian Living › Christian Inspirational

Paperback ISBN: 979-8-89307-018-7
Hardback ISBN: 979-8-89307-019-4
Audiobook ISBN: 979-8-89307-020-0
eBook ISBN: 979-8-89307-017-0
Library of Congress Control Number: 2024901702

I dedicate this book to the Almighty God who has been the source of my strength, grace, and wisdom throughout my life.

To my awesome parents, Donald Sr. and Mary Bethune, who have planted the seed of knowledge in my mind and nurtured it throughout my life.

And to my wife, Dashika, son Jeremiah, and daughter Jo'mariah Bethune, who have walked every step of this wonderful journey with me.

TABLE OF CONTENTS

Introduction 7
Part I: The First Six Months 9
Part II: The Second Half 67
Part III: Going Home 101
Conclusion 107

INTRODUCTION

Uncertainty weighed over my first deployment to Iraq, casting a shadow that made each day feel unpredictable. This was more than simply going to battle; it was also about stepping into the unknown and spending my first time apart from my wife. It was difficult, and to be honest, I'm not sure how we would have survived if it weren't for my faith and my determination.

The first year I was deployed was a challenge, not just on the battlefield but also in my personal life. It was more difficult than I could have ever imagined being apart from my wife, both physically and mentally. However, my faith served as a rock in the middle of all the chaos, giving me courage when things became hard. And fortunately, my deployment also helped me reflect on my faith build a closer relationship with the Lord.

This book is more than just war stories, it's also about the real struggles of being a soldier and a husband. I am also sharing this so it can serve as a form of encouragement and motivation for anyone who might read this.

I want you to see the ups and downs, not just in the deserts of Iraq, but in the ups and downs of my heart. I share this to inspire rather than to boast. These pages may offer you some optimism, a reminder that we can muster the will to persevere in the face of conflict and distance. Maybe you'll find a bit of hope in these pages, a reminder that even in the midst of war and distance, we can find the strength to keep going.

After each part of the book, there is a prayer, and I hope that these will help you reflect on God's Word and will inspire you as well. We always need God's guidance, and hopefully these prayers will help you become closer to God.

So, come join me on this journey, where the battles aren't just out there, but inside, where we find out what we're really made of. I hope you enjoy this book, and may it serve as both an inspiration and a blessing for those who read it.

- Ronald Bethune

PART I: THE FIRST SIX MONTHS

This part serves as a compelling start to my deployment, revealing the emotional landscape of my journey, which was marked by uncertainty, initial struggles, and a gradual change in my physical and spiritual strength, as well as my resilience.

MY FIRST DEPLOYMENT

First Day
April 13, 2003

Dashika (my wife) and I woke up early since our plan was to attend the service at Zoe Christian Center in Lawton, Oklahoma before I set off for my deployment to Iraq. After a final shared breakfast, I suited up in my military uniform, and together, we made our way to church. The Pastor's heartfelt prayers embraced both of us during the worship service. Come afternoon, just before roll call, we assembled at the unit to receive the gear for my deployment.

Picture of the sleeping tent for all our gear.

Temporary Stay
April 14, 2003

Upon landing in Kuwait on a spacious two-deck 747, we settled at Camp Wolf near Kuwait International Airport. This camp served as a staging area for US troops heading north into Iraq during

PART I: THE FIRST SIX MONTHS

Operation Iraqi Freedom. Our stay there was temporary, awaiting the call to move to a new location. The weather was pleasant during this time. At 6:00 pm, my comrades and I headed to the DFAC, or the eating facility. For dinner, I had a bottle of water, along with vegetables, hash potatoes, and pepper steak. Despite the meal, my thoughts lingered on my beloved wife, the person I cherish the most.

Camp Activities
April 15, 2003

After a short night's sleep, waking up at 5:00 am for breakfast felt a bit challenging, especially since I had only gone to bed around 4:00 am. Fortunately, I caught some rest on the plane from Frankfurt, Germany, to Kuwait. Breakfast at the spacious Camp Virginia included sausage, biscuits, croissants, pineapple, and orange juice. The camp was bustling with different activities—some played spades, others wrote letters, and some simply took a nap. While others went about their business, I was on a mission to find the mailing address to receive heartwarming letters from my wife. By 1:00 pm, lunch was served, but the long walk to the dining area made it less enjoyable. The humidity was intense, turning the chow hall into a sauna, making everyone sweat as soon as they stepped inside. After lunch, I decided to take a short nap.

Tough Weather
April 16, 2003

As the strong wind and swirling dust surrounded us between 7:00 am and 9:00 am, the weather got really tough. The sand whipped around like crazy, making our faces go all white, and we could barely see anything—just about 10 feet ahead. The sandstorm was super strong, and it seemed as if everything turned into a snowy day, all covered in white. After going through the storm, my skin, which used to be smooth and dark, now looked sandy-dark. It was like the storm left a mark on me, changing how I looked after being out there in the wild weather.

Long Day Ahead
April 19, 2003

The day has been going well so far. On Thursday, I had an early start at 4:30 am, heading to Kuwait harbor to pick up some Humvees. It turned into a long day, and I didn't get back to bed until Friday morning around 11:30 am. After a late-night shower at 12:30 am, the day finally came to an end. Today, I kicked off the day with a hearty breakfast. Given the dusty and sandy conditions in the air, we've been diligent about cleaning our weapons 2 to 3 times a day. As I reflect on what I have been doing these past few days, it feels like a good time to read the Bible, write a letter to my wife, and unwind with some comforting gospel music. These moments help me relax and stay connected to my faith.

All of our gear we left outside was untouched.

PART I: THE FIRST SIX MONTHS

Connecting with Family
April 20, 2003

Starting the day with breakfast and connecting with my wife over the phone was truly uplifting. Though I couldn't see her, the sound of her voice brought a sense of calm to my mind. I also had the chance to chat with my mom, Mary, and my sister, Lashannon, making the morning even brighter. The only downside is the limited time for phone calls, we get just 15 minutes, and I wish there was more time to catch up. Looking ahead, our Personnel Supervisor has informed us that tomorrow we'll be taking a pill to prevent malaria, emphasizing the importance of staying proactive about our health.

Quiet Bond
April 21, 2003

I started my day early, waking up at 1:50 am. After a refreshing shower, I decided to play some gospel CDs. What caught my attention was that almost everyone around here wears a spiritual cross. It struck me how, despite our different backgrounds, this simple symbol created a quiet bond among us. It's like a shared understanding that we're all in this together. As I lay there in the dim light, enveloped by the comforting sounds of gospel music, it felt like more than just melodies. It was a reassuring connection, a little source of strength in the midst of the unknown.

> It struck me how, despite our different backgrounds, this simple symbol created a quiet bond among us. It's like a shared understanding that we're all in this together.

MY FIRST DEPLOYMENT

Sandstorms and Snakes
April 22, 2003

We just had a bad sandstorm that lasted for a while last night and a snake was in the Bravo Unit/Organization tent. When the storm was over, the floor was covered in sand and everyone's gear had sand all over it. We had to walk a half mile to chow. The phone line usually takes about 1 to 2 hours and the PX is the same—we wait in line to get in the door. It's 9:45 am and we are still waiting for the bus to take us from Camp Virginia to Kuwait City port to pick up some vehicles. The bus was to show up at 7:00 am but we are still waiting.

Me posing inside the B-Hut.

Scorching Temperatures
April 23, 2003

Last night was quite an adventure as we headed out late to pick up some vehicles, finally returning to camp around 4:00 am. I was so tired that I couldn't help but doze off whenever we made a stop, but I'm thankful we all made it back safely. Now, it's 11:00 am, and it's starting to get pretty hot inside our tent. Oddly enough, it's cooler outside than in here.

PART I: THE FIRST SIX MONTHS

Speaking of our surroundings, the port in Kuwait is quite a sight with a massive oil refinery hosting enormous ships. At the chow hall, my lunch consisted of beans, orange juice, fries, and two cheeseburgers. It was so warm that even the cheese on my burger started to melt. The temperature today has reached 101°F, and that's not even accounting for the heat inside the tent. Despite the scorching weather, each day brings a new set of experiences and challenges to navigate.

God is in Control
April 24, 2003

Today, we had a class on staying safe in the bustling city of Baghdad, Iraq, courtesy of the 3rd Military Police (MP) unit from Fort Stewart, Georgia. It was a really useful session, but I found myself a bit disappointed that my driver is going instead of me. But, you know, things happen for a reason. Perhaps God thinks I'm not quite ready, physically or mentally, or maybe both. I trust that God's got control over everything.

My desire is to go, take pictures, witness things I've never seen before, and share these experiences with others. Now, it's 1:00 pm, and I just finished doing laundry. It's scorching outside, so I brought most of it back in, and it surprisingly took about 2 hours to dry due to the intense heat. After sorting my laundry, one of the Non-Commissioned Officers (NCOs) gave me a haircut, and they joked that I now resemble "Michael Jordan." It's little moments of fun and casual conversations that add some humor and camaraderie to our days here.

MY FIRST DEPLOYMENT

The Al-Faw Palace in Baghdad, Iraq which contains 62 rooms and 29 bathrooms.

> It's little moments of fun and casual conversations that add some humor and camaraderie to our days here.

Challenges and Conversations
April 25, 2003

It's 10:41 am, and the heat is growing significantly. I find myself sitting, leaning on my bags, lost in thoughts about the special woman I had to leave behind, the one I love so much. Conversation hasn't been my strong suit since our arrival, yet, amid the challenges, I hold on to a positive outlook. But love and my wife keeps me

PART I: THE FIRST SIX MONTHS

strong, even in this unfamiliar place. It's like a warm light guiding me, giving me hope and determination, to keep going.

Risking Trouble
April 26, 2003

This morning started with me getting up early around 5:45 am. I headed to use the phone and tried calling my wife from the internet café, but she didn't pick up. Instead, I had a chat with my mom. Unfortunately, when I finally got through to my wife, I only had a short time left, and it made me a bit emotional because I couldn't talk to her as much as I wanted. Still determined to connect, I got back in line, even if it meant risking trouble for coming back late.

Then I had to be ready at 1:00 pm to catch a bus to Kuwait for some car pickups. I saw some camels on the way, and on the way back, there was so much to see—sheep, tents scattered throughout the desert where people live, and even an oil plant. The most unexpected sight was Kuwait City in the middle of the desert, which seemed insane to me. The day was not without its obstacles, but it also provided moments of connection and awe.

Me standing beside a camel while it was resting.

Finding Strength and Peace
April 27, 2003

Today has been good so far. I had a chat with my wife, and last night, we faced a pretty bad sandstorm. It turned into rain, but the dust didn't completely settle. The strong winds even knocked over a couple of portable restrooms, and one tent just collapsed. When we went to the chow hall, it was packed with about 10,000 soldiers. The camp is supposed to hold 6,500, so there were around 4,000 extra soldiers. It was quite crowded.

One of my fellow soldiers shared something interesting. He mentioned that since he started reading the Bible every day, he's found a way to relax. He said he learned it from me, that reading the Bible does help. It's moments like these that show the importance of finding strength and peace in simple things, even in the midst of chaotic situations. I suggested he read Psalm 23 (NLT), a passage that has brought comfort to many (including myself) in challenging times:

The Lord is my shepherd;
  I have all that I need.
He lets me rest in green meadows;
  he leads me beside peaceful streams.
He renews my strength.
He guides me along right paths,
  bringing honor to his name.
Even when I walk
  through the darkest valley,[a]
I will not be afraid,
  for you are close beside me.
Your rod and your staff
  protect and comfort me.
You prepare a feast for me
  in the presence of my enemies.

PART I: THE FIRST SIX MONTHS

You honor me by anointing my head with oil.
My cup overflows with blessings.
Surely your goodness and unfailing love will pursue me
all the days of my life,
and I will live in the house of the Lord
forever.

Maintenance Checks
April 28, 2003

Today began with a hectic morning full of tasks. First, I went with a soldier to go over his finances and make sure everything was in order. After that, I set aside time to perform Preventive Maintenance Checks and Services on my vehicle to ensure it is in excellent shape.

During the day, I noticed an issue with the trailer tire—it wouldn't hold air due to a stubborn piece of metal lodged in it. As lunchtime rolled around, instead of taking a break, I found myself changing the tire. Thankfully, my platoon sergeant pitched in to help, turning what could have been a challenging situation into a collaborative effort. Despite the unexpected tire hiccup, the day's tasks were tackled with teamwork and efficiency.

Evaluation Report
April 29, 2003

Things are going well, so far. Right now, my platoon sergeant is working on my NCOERs because my annual is due by May 1, 2003. These evaluation reports are crucial—they're the main source of info for decisions about NCO personnel management. They act as a guide on how well I'm performing, helping in my development, and making sure I'm contributing to the organization's mission. Essentially, they provide extra info for everyone involved in the rating process. Hoping all goes well, and the reports reflect positively on my performance.

Simple Routine
April 30, 2003

Today has been great so far—I woke up feeling good. For the past two days, I've been riding back and forth in the Humvee with the guys, making rounds to their guard points. As my thoughts turn to my wife, I observe the surroundings, and I can't help but notice that the air outside is filled with dust, creating a hazy atmosphere. I am looking forward to some peace and quiet, first immersing myself in my Bible, and then in the songs on my faithful CD player. It's a simple routine, but it's my go-to for finding peace and a comforting connection in the midst of the dust and heat.

Dusty Conditions
May 1, 2003

I just had a wonderful conversation with my wife, and it brings me joy to know she received my first letter. I sent it on the 15th of April, and she got it on the 30th of April—roughly three weeks later. Currently, it's 12:45 pm, and we've been dealing with a sandstorm since 8:00 am. Stepping into the tent, everything is coated in a layer of dust, sand, or maybe both. Despite the dusty conditions, my next plan is to write another letter to the most important woman in my life, my beloved wife, whom I hold so dearly in my heart.

PART I: THE FIRST SIX MONTHS

Early to Bed
May 2, 2003

My section chief met me later in the day to make revisions on my NCOER after I completed the unusual laundry duty with the 5-gallon bucket and scrubboard. It's only 9:00 pm, but I'm already getting ready for bed, marking the first time this week that I can genuinely relax and catch up on some much-needed rest. From laundry challenges to NCOER adjustments, the simplicity of this routine makes sleep a pleasant moment of peace in the midst of a hectic week.

Comforting Connection
May 5, 2003

It was still another day of scorching heat. As I walked, sweat was literally rolling down my body. I actually discovered a little hack for the bathroom—it's best to go early, before 8:00 am, or late in the afternoon when the sun is easing down, and it's a bit cooler. And let me tell you, sitting in the tent? You can feel the heat just pressing in from every side.

Later in the evening, roughly around 7:00 pm, I had a heartfelt conversation with my wife. The joy in her voice when she heard mine was truly uplifting, it absolutely made her day. And you know what? It made mine too—a comforting connection in the midst of the heat and challenges of the day.

Moments of Relaxation
May 6, 2003

I started the day feeling great. Now, I'm just chilling, trying to keep cool with this heat settling in. In the midst of the scorching weather, I'm taking it easy and making the most of the day. Despite the blazing sun, these quiet moments of relaxation provide a sense of peace. I'm simply lying here, figuring out little ways to stay cool and beat the intense heat. When the temperatures soar like this, not much else is happening—just me, the heat, and finding simple pleasures in the stillness of the day.

Familiar Face
May 7, 2003

In the morning, I started the day feeling fine, but as the clock struck 8:00 am, the heat intensified, and beads of sweat began to drip down my forehead, making the tent feel warmer, so it felt like your body was on fire. Even with the scorching heat, I found myself engrossed in studying for the board. Unexpectedly, I glanced across the door leading to another tent and saw a face that resembled Brother Lowe—and indeed, it was him! The shared excitement of recognizing a familiar face from church made me feel that the day was much better regardless of the heat. By 1:00 pm, the weather takes an unexpected turn with rain, yet the air remains filled with dust, creating a unique blend of elements outside.

A Pleasant Surprise
May 8, 2003

I began the day by waking up early, finding solace in reading the Bible, and taking a refreshing shower at 3:00 am. As the sergeant of the guard, my first task was to check on the guards at the motor pool, ensuring everything was in order. Following that, I decided to tackle the mundane yet necessary task of doing laundry, a routine that brings a sense of normalcy to deployment life. With these chores behind me, I indulged in the soothing melodies of my CD collection.

PART I: THE FIRST SIX MONTHS

By 6:00 pm, a pleasant surprise awaited us in the form of a massive 34-inch colored TV with satellite access. The thought of having entertainment lifted everyone's spirits, and a collective effort unfolded as people eagerly pitched in to set up the TV in the SUV, creating a shared space for relaxation and camaraderie.

Beating the Heat
May 9, 2003

Today was another hot day. The kind where the sun makes everything feel sticky, leaving wet spots when you rest your arms on your legs. The heat stuck around all day, making everything humid. Around 8:00 am, we decided to beat the heat by watching TV together. It turned the day into a fun break, and we kept watching until about 10:00 pm. During this time, I squeezed in my second haircut at 5:00 pm, a little self-care in the middle of our routine.

Surprisingly, the weather wasn't as blazing as yesterday. Even though it was still hot, it felt a bit better. The day had its own mix of things to do and some moments to relax, giving a bit of variety to our usual deployment routine.

Comfort and Consistency
May 10, 2003

I had a heartfelt talk with my beautiful wife, whose overwhelming beauty never fails to captivate my eyes. The joy of hearing her voice brought warmth to my day. Despite the weather's unpredictable mix of rain and dust, it remained warm outside, creating an unusual combination.

I find it amusing because most of the time, I'm the first one up and dressed, a routine that adds a touch of consistency to the unpredictable surroundings. There's a sense of comfort and familiarity in these moments of connection and strange weather, a reminder of the constants that anchor us in the midst of the deployment experience.

Desolation and Disorder
May 11, 2003

Here at Camp Virginia in Kuwait, it's Mother's Day, although it's not the day yet in the United States. Around 6:30 am, the heat started to kick in, making me break a sweat. On this special day, I want to express my deep love for my mother. However, the TV situation is becoming a bit troublesome, with trash left in the area, the overly loud volume, and its proximity to my sleeping cot. Today, two individuals wanted a drastic haircut, opting to go for a complete shave.

During informal conversations with fellow soldiers who had the experience of traveling through Baghdad, they vividly portrayed the towns they encountered as bearing a striking resemblance to junkyards. The details they shared painted a picture of a landscape cluttered with debris and discarded items, evoking a sense of desolation and disorder. Their reports provided insight into the difficult conditions and aftermath of conflict in those locations, emphasizing the visible influence on the local environment.

God's Blessings
May 12, 2003

God's blessings have accompanied me on this deployment journey. I learned that there were widespread tire issues due to cracking and dry-rotting, causing blowouts left and right since they left Fort Sill, Oklahoma. The challenges with the tires added a layer of difficulty to our journey.

PART I: THE FIRST SIX MONTHS

Despite the obstacles, I felt a twinge of regret for not being able to connect with my mom to wish her a Happy Mother's Day. However, I find solace in the understanding that she comprehends the challenges and limitations of communication in these circumstances. The difficulties in communicating with loved ones serve as a reminder of the hardships of deployment and the sacrifices made by soldiers and their families.

> The difficulties in communicating with loved ones serve as a reminder of the hardships of deployment and the sacrifices made by soldiers and their families.

Good and Bad News
May 13, 2003

This morning, around 4:00 am, I had a conversation with my mom and dad. I wished my mom a Happy Mother's Day, but the joy of the moment was dampened by the heartbreaking news from my dad. He shared the devastating news that one of my classmates/friends was killed by a drunk driver in Sumter, South Carolina. The incident cast a shadow over what should have been a joyous event.

Later, I connected with my lovely wife, and hearing her voice provided a comforting contrast to the sad news. As I interacted with my fellow soldiers, I learned that most of the guys from our unit had returned to Camp Virginia. However, there was uncertainty about their stay here, especially given the constant arrival of people returning from Iraq. The ebb and flow of emotions, from the sadness due to loss, to the warmth of connection, emphasized the overall experience of being deployed.

A Mix of Experiences
May 14, 2003

Today brought a mix of experiences. I made a unique purchase—a 25 Iraqi Dinar (IQD) from Iraq for $1. This currency, once valued, now holds little worth due to the fall of Saddam Hussein. Interestingly, a vendor tried to sell it to me for $5, but I declined, recognizing its diminished value.

In the midst of this transaction, I found solace in my spiritual pursuits. I spent some time reading the Bible, allowing music from my CD player to accompany this reflective moment. The day unfolded with its particular blend of business conversations and spiritual moments, emphasizing the contrasts that accompany the deployment experience.

Reflection
May 15, 2003

This morning, I woke up drenched in sweat, even though all I did was lie on top of my sleeping bag. It felt like someone had poured water on me. As I waited in line for the phone, the intense heat made me doze off twice. This was around 10:00 am.

In the afternoon, I took a stroll to what we call the "dusty room," which is the gym. After a workout, I headed to our old pad, now occupied by someone else. While listening to music, I found a quiet spot to sit and reflect on the Lord. However, despite the personal time, we weren't receiving any mail. The delay was due to our mail shuttling back and forth between Iraq and Kuwait.

PART I: THE FIRST SIX MONTHS

Waiting for Mail
May 16, 2003

The routine remains unchanged, with anticipation for mail from my loved ones adding a touch of eagerness to each day. Unfortunately, the mailbox remains empty so far. On a positive note, I had a reassuring conversation with one of the commanders from Germany who shared that everything is okay and there are no issues.

A Comforting Moment
May 17, 2003

This morning, the scene that greeted me was sand, coating not only me but also my gear. Even with windows in our temporary tent, the relentless sand seemed to find its way in, creating a constant, fine rain indoors. By 1:00 pm, I seized the opportunity to talk to my beautiful mom, dad, and wife. The reassurance of their well-being is always a comforting moment in this experience. At 10:00 pm, I'm getting ready for bed, expecting to say my prayers and immerse myself in a quiet reading of the Bible.

Good Energy, Good Vibes
May 18, 2003

This morning was filled with good energy and good vibes as I just finished talking to my hospitable and courteous lady. Later, at 11:00 am, I attended a worship service that concluded at 12:35 pm. The lesson imparted was a reminder not to let the desert surroundings get us down but to keep our heads up. The preacher emphasized that one day, willingly or not, we will all bow down and confess to God.

With all this heat and humidity, I spotted a woman lying in the sun, attempting to get a tan in temperatures exceeding 100°F.

Around 5:30 pm, a severe sandstorm hit, causing the DFAC beams in the tent to shake. The left side of the sky grew dark and dusty, while the right remained clear as crystal. By 9:30 pm, I've just wrapped up watching the movie "8 Mile," which was pretty good. Next on the agenda is delving into the study of the Bible.

Evening Inconveniences
May 20, 2003

Last night, around 6:00 pm, our tent lost electricity, and for a while, it got a bit hot. Fortunately, a breeze came through, making it bearable enough to sleep. I'm a bit disappointed that I still haven't received the mail my wife sent on the 3rd of May, while others have. It's a small frustration amid the bigger challenges.

Taking a shower became a bit of a struggle last night. I had to try two different places before I could find a working shower. Apparently, there was a water shortage, and people were taking too long in the showers. Despite these inconveniences, I had dinner with a NCO who recently retired in June. Another NCO, who used to be the senior NCO in the DFAC in Germany, is now here and plans to settle in Georgia after retirement. It's interesting how connections from different parts of our military journey continue even in unexpected places.

Intense Heat
May 21, 20023

It's currently 12:38 am, and the temperature is around 111 degrees Fahrenheit. The heat is so intense that the air conditioning in our tent provides only minimal relief. Despite the challenging conditions, I continue to read the Bible and study for the board. The power was restored around 1:45 am, bringing some respite.

PART I: THE FIRST SIX MONTHS

Today, I came across a paper stating that between June and September, temperatures can reach 100°F to 120°F by 10:00 am. The extreme heat is a constant presence. Interestingly, a man driving a water truck gave me a half Kuwaiti KWD, a small but unexpected gesture that adds a touch of positivity to the day.

Fireworks Display
May 22, 2003

The temperature has been constantly increasing, signaling the onset of another hot day. Last night, Explosive Ordnance Disposal was detonating ammunition, creating a spectacular display of big mushroom cloud, complete with vibrant flashes of red and orange. Despite the nighttime pyrotechnics, I managed to talk to my wife, my queen, who's doing well.

In the early hours of the morning, around 2:00 am, I took advantage of the quieter time to shower, avoiding the usual crowds. I prefer solitude when it comes to both showering and sleeping, especially if I can position myself close to the exit. Today brought a special joy as I received my first letter from my wife, sent on the 23rd of April. The excitement is palpable, and it seems like a Christmas morning for me and the other guys who share in the joy.

Unfortunate Incidents
May 23, 2003

As the clock approached 11:00 am, the relentless heat persisted. I spent the entire morning, starting at 7:00 am, walking and engaging in various activities. After a quick breakfast, I participated in a Police Call, picking up trash to maintain cleanliness. My time was further occupied by walking back and forth to assist a staff sergeant in obtaining Morale, Welfare, and Recreation equipment for the soldiers in Iraq.

MY FIRST DEPLOYMENT

An unfortunate incident unfolded as we discovered the Conex, left unlocked, had been robbed. Our unit, in collaboration with the Bravo and Charlie Units, returned for supplies only to find that someone had taken advantage of the situation. Additionally, they shared the unsettling news of coming across a deceased individual on the side of the road during their journey.

Evident Happiness
May 24, 2023

I started my day at 4:00 am, feeling energized as I took on the task of doing laundry. Connecting with my father and wife brought joy, their happiness evident in our conversation. The night before, I had prayed for the necessary materials to study for the promotion board, specifically to advance to first sergeant.

A chance encounter occurred when a first sergeant from the Charlie Unit entered the tent. I took advantage of the situation and asked about the board subjects, and to my relief, he confirmed that he had them. Grateful for the timely assistance, I couldn't help but attribute it to the Lord's grace.

Blessed Day
May 25, 2003

Approaching 8:00 am, I relished another blessed day the Lord has granted me. Sundays afford us the luxury of sleeping in a bit, a small respite in our routine. Attending a USO concert at the camp started as an enjoyable experience, with music filling the air. However, things took an unexpected turn when some individuals tampered with glow sticks, resulting in a colorful, albeit, messy, display. The military police promptly intervened to halt the impromptu light show.

PART I: THE FIRST SIX MONTHS

At 1:00 pm, I returned from both the church service and the dining facility. The chaplain's words lingered in my mind, emphasizing that being out here in the deployment could foster a closer connection to the Lord. This simple message struck a chord with me, presenting a perspective that surpassed the challenges of our environment.

Interesting Conversation
May 26, 2003

Today marks another beautiful Memorial Day, and I find myself in good spirits. My only wish is for the care package my wife sent to arrive—the anticipation adds a touch of excitement to the day. An NCO remarked yesterday that my routine involves nothing more than studying and sleeping. I disagreed, asserting that everyone is entitled to their opinion. The exchange ended with him expressing a lack of interest in further discussion. Despite that experience with an NCO, I continued to engage with members of other units, curious to learn about their origins and experiences.

Third Marriage Anniversary
May 27, 2003

It's 4:00 pm, and I've just got off trash-burning, lasting only an hour. Today holds special significance as it marks the third anniversary of my marriage to my lovely wife. Reflecting on our journey, we've experienced both ups and downs, but the positive moments have consistently outweighed the challenges. I'm grateful to the Lord for keeping us together this far. Despite the distance, I wish we could celebrate our love in person, but I look forward to making up for lost time when I return.

MY FIRST DEPLOYMENT

Dinner was quick, taking only 15 minutes from walking in, eating, then leaving. On the menu were a quarter of a leg, string beans, rice, three slices of watermelon, and three small pineapple juices. The constant sound of whirling sand drew my attention as I stepped outside. The specks seemed to pervade the air, making me feel an uncomfortable dry sensation on the skin.

Receiving my First Care Package
May 28, 2003

Today was a good day, as I ventured to Camp Doha for the first time. It was a brief Rest and Recuperation, just for one day, but a welcomed break. The highlight of my day was receiving my first care package, dispatched on the 2nd of May. We can somehow compare the sheer joy of finally getting it to the excitement of obtaining your driver's license—smiling from ear-to-ear and feeling elated.

During my time in Camp Doha, I encountered a sight to behold—numerous camels and a man, presumably a camel herder. The Kuwaiti landscape showcased impressive houses, and I observed women donning traditional attire, covering themselves entirely, with only their white eyes visible. It was a fascinating glimpse into the local culture.

Man standing outside a village in Iraq.

PART I: THE FIRST SIX MONTHS

Health Improvement
May 29, 2003

This was a good day, notwithstanding the relentless heat outside. Staying cool is a challenge, especially during lunch when the tent's AC struggles, freezing up, and blowing warm air. To boost my overall health, I decided to take a ginseng pill for the first time. I heard of the benefits of ginseng, and I wanted to try it. Now, I'm attempting to get some rest before diving into my study session.

After my study session, I took a short nap. I woke up feeling better, though the heat tends to induce sleepiness. Interestingly, I had a dream, but the details slipped away from my memory upon waking. It seems the heat's influence extends even into the realm of dreams.

A Reminder of Home
May 30, 2003

Talking to my wife, even from far away, makes me feel normal in this place. Each chat keeps me connected to the familiar feeling of our relationship, even though everything else is strange. Now it's 8:00 pm, and the day is almost over. I just had a simple and tasty dinner. My wife's picture is a reminder of home and makes me feel comforted and at peace. I usually kiss it before I go to sleep.

As night fell, the silence around my tent was perfect for reading the Bible. It's 9:00 pm, and with the soft light inside the tent, I read and found comfort in my faith. Now, as the day comes to a close, I'm ready to sleep, grateful for the little things that make each day easier here.

MY FIRST DEPLOYMENT

Simple yet Meaningful
May 31, 2003

Saturdays are usually quiet here, not much going on. I did a bit of studying and then took a nap. In the evening, I dedicated some time to letter writing, writing two heartfelt letters. Later, I headed to the dining area for a meal. As part of my daily routine, I took a stroll, seizing the opportunity to collect various foreign currencies. And with that, my day came to a close, marked by these simple yet meaningful activities.

Unexpected Encounters
June 3, 2003

Today, I visited Camp Doha and had some unexpected encounters. I bumped into an NCO from our church who happens to play the guitar. Later, I ran into another NCO, a familiar face from the post office back in Germany. During our conversation, we unexpectedly crossed paths with the dining facility manager at Kuwait City Airport, adding a touch of familiarity to the day.

Unforeseen Challenge
June 4, 2003

Today presented an unforeseen challenge as I talked to our rear detachment NCO and learned that my sleeping cot had been given away, leaving me without it. Now, I find myself on a mission to locate it. Losing personal belongings in such scorching conditions adds an extra layer of challenge to an already demanding environment. To add to the day's difficulties, the temperature soared to a scorching 130°F.

PART I: THE FIRST SIX MONTHS

Small Misunderstanding
June 5, 2003

Today, there was a slight mishap, but by the end of the day it became better, upon seeking medical attention for what I thought might be a strep throat, it turns out it's merely a sore throat. I felt a slight relief of sorts. In the midst of the day's activities, I had a heartfelt talk with my wife, only later getting caught in a tiny misunderstanding. Despite the emotional pressure and a few tears, we are determined to keep our bond strong even when distance and challenges try to test us.

Silver Lining
June 6, 2023

Enduring escort duty from 6:00 pm the night before until 6:00 am today, my relief didn't come until around 8:00 am. The task included recharging many generators alongside a fellow soldier. When I returned to my post, desperate for a replacement, I had to embark on a hot trek on foot, a walk that intensified a sense of irritation, made worse by the oppressive heat. I was irritated by the lack of transit. However, a silver lining appeared when some compassionate comrades later picked me up, relieving the frustration that had briefly clouded my day.

Conquering Challenges
June 7, 2003

Saturday mornings here follow a predictable routine, marked by the post police call—cleaning up the camp and picking up scattered trash. Unfortunately, the anticipated arrival of mail has yet to become a reality for me. It can be discouraging at times, but I believe that with God on my side, I can conquer every challenge that comes my way. The weekly housekeeping provides some order in the middle of the unpredictability, and despite the lack of mail, the optimism and belief in divine assistance keep my mood bright.

Meaningful Conversations
June 8, 2003

I had a meaningful conversation with my wife today, and then I attended a church service that left a positive impact. The discussions at church revolved around the health challenges faced by several young children, bringing a sense of communal concern. One particular conversation struck a chord with me—the young man whose pregnant wife is grappling with various struggles, notably a severe risk of bleeding to death due to clotting issues. Touched by this, I dedicated a day of fasting, praying for a swift return home, fervently wishing it won't extend until October. Unfortunately, the mailman informed me that forwarding mail is now impossible. The temperature rose to 115°F amid these emotional and logistical difficulties.

Hoping to be Reunited
June 9, 2003

Early this morning, at around 2:45 am, I rose to do my laundry before engaging in a comforting conversation with my wife. The sacrifice of sleep has become a routine for me since arriving here, a small price to pay for the solace her voice brings. Despite the challenges, I hold a deep belief in my heart that I'll be reunited with her by August this year. Amidst the daily trials, I continue my commitment to reading the Bible, seeking understanding, and praying for a closer connection with God, finding strength in faith during these testing times.

PART I: THE FIRST SIX MONTHS

Compassion Over Negativity
June 10, 2003

    Returning from Camp Doha, I found myself among sleeping comrades, standing out as the only one fully dressed. It's a quiet moment, and I realize someone has swept a considerable amount of dirt near my sleeping cot, thoughtlessly leaving it behind. Positioned at the end of the doorway, it seems they deemed it acceptable. In my attempt to muster patience, I hope the responsible party would return to rectify the situation, yet, regrettably, no one came. Despite this, I chose to give them the benefit of the doubt, opting for compassion over negative thoughts.

Me standing in front of our bunk beds.

MY FIRST DEPLOYMENT

Remaining Resilient
June 11, 2003

I just wrapped up a conversation with my precious wife. She mentioned that at the family readiness group (FRG), news circulated that a senior personnel email hinted at a potential homecoming by Thanksgiving. However, I'm steadfast in my belief, rooted in both mind and heart, that we will reunite by the 15th of August. My faith in God assures me that we'll be home even sooner, as He holds the ultimate authority. While my wife may be a bit disheartened, we remain resilient with God by our side.

On the other hand, the temperature today is expected to reach 118°F or 48°C, but around noon, it spiked to about 137°F or 140°F. I also received a heartfelt package from my mom, which was sent on the 7th of May, lifting my spirits—fortunately it just took 36 days for it to arrive.

Jet Magazine
June 12, 2003

In Kuwait, it was another scorching day. I then had a sentimental talk with my wife, declaring her my lady of the year for 2004, and proudly displaying her in my own Jet magazine. On the bright side, I was able to secure a sleeping cot, despite a switch. Despite having a new one, they apparently decided to swap it for another.

Mailing Issues
June 13, 2003

Navigating another day, learning it's Friday the 13th, with tomorrow celebrating the Army's venerable 228th birth anniversary. However, the passing of time hasn't led to any letters from my mother, a situation worsened by a mail system issue. Nonetheless, I am hopeful that things will improve, especially now that I have shared my new address with my loved ones. Due to a mix-up among the guys here, the mail intended for the camp down south usually end up heading north.

Running and Reflecting
June 14, 2003

I woke up early and prepared to run. I started at 5:00 am and finished a solid 2.7-mile run, maybe even 3 miles after a while. Running through the soft sand was tougher than expected, but I kept going without stopping.

After a vigorous run, I took a moment to gather my thoughts, calming myself to write about what I want to share, talking about developments and pouring my heart out in a letter to my family. I'm crossing my fingers that the letter reaches them within the next 14 days, delivering a sense of connectedness and warmth across the distance.

Timely Reminder
June 15, 2003

Today is Father's Day. I count my blessings for having a living father whom I can cherish and know. A heartwarming conversation with my family assured me of their well-being. Attending the church service added positivity to the day, affirming a crucial life lesson: the importance of patience. It served as a timely reminder that, even in challenging circumstances, maintaining a patient and hopeful outlook can contribute significantly to one's well-being.

MY FIRST DEPLOYMENT

> *. . . even in challenging circumstances, maintaining a patient and hopeful outlook can contribute significantly to one's well-being.*

A somber statement from an NCO highlighted the grim reality for soldiers in Iraq, where the possibility of injury or loss remains every day. This Father's Day serves as a poignant reminder of the sacrifices made by those in service.

Sudden Change
June 17, 2003

Today unfolded as another scorching day, prompting a morale trip to Camp Doha. While the experience was relatively standard, there was a unique sight as we departed—a tractor-trailer transporting a load of camels, adding a touch of fascination to the journey. Curious about the sudden change, I find myself facing duty five times this week, unlike my usual routine. Nonetheless, I intend to complete the tasks at hand.

Mom's First Letter
June 18, 2003

As the sun blazes with a scorching 117°F, the relentless heat continues to take its toll, resulting in daily incidents of heat injuries among our ranks. Amidst the challenges, a glimmer of relief came as I finally received the first letter from my mom, a welcome connection to home. In the midst of the relentless heat, the joy of receiving my first letter from my mother outweighs the struggles of the scorching heat.

PART I: THE FIRST SIX MONTHS

Blessed Opportunity
June 19, 2003

Blessed with the opportunity to talk to my wife, even as she intermittently drifted into sleep during our conversation, I reflected on the frequency of our communication. Unlike many soldiers who connect with their families only once every 2 or 3 weeks, I consider myself fortunate. Additionally, a small but significant source of joy was the arrival of my first letter with the new address, taking just 8 days to reach me.

Joy Night
June 20, 2003

I had the honor of attending Joy Night on this unforgettable day filled with memorable events. It is a special service fostering fellowship, worship, and a profound sense of refuge. The experience proved to be deeply moving, evoking emotions so powerful that my eyes couldn't help but get wet with unshed tears, making me speechless.

Future Home
June 21, 2003

Today begins with good news, as my wife informs me that the housing office has a prospective house for us to look at. Despite our physical distance, we fast and pray, praying for guidance in deciding on the best choice for our future home, with a particular need for a drive-in carport. Meanwhile, a fellow soldier notices my breadth of expertise during a routine police call, commenting that I appear knowledgeable and experienced.

MY FIRST DEPLOYMENT

Faith and Personal Beliefs
June 22, 2003

I woke up feeling positive this morning and attended church, which was alright. I had a conversation with my wife, and she's doing well. Upon my return from church, my Non-Commissioned Officer in Charge (NCOIC) asked me if I had found Jesus, prompting a brief but meaningful conversation about faith and personal beliefs. I responded by sharing that Jesus is within everyone, and knowing Him is a personal experience. After that, he didn't say much and went to lunch. Additionally, I consider myself a good person, for instance, when the water in the cooler runs low or runs out, I take the initiative to refill it. I believe that the Lord will bless me for my actions.

> ... Jesus is within everyone, and knowing Him is a personal experience.

Memorable Encouragement
June 23, 2003

Today turned out to be a good day, and the workload wasn't too demanding. The heartfelt encouragement card from my family created a warm connection, lifting my spirits and setting a positive tone for the day. In the later part of the day, I had a pleasant and uplifting conversation with them over the phone, further contributing to the overall positivity and making the day memorable.

PART I: THE FIRST SIX MONTHS

A Sense of Peace
June 24, 2003

The extreme temperature caught me by surprise, emphasizing the importance of staying cool in this environment. However, having the air conditioner on made a big difference in this sweltering heat.

Fortunately, I could speak with my wife, but she's a bit sad because I'm away. I reassured her that everything will be okay and that I love her. As I reflected on the words of Psalm 91 (NLT), it brought a sense of peace amidst the challenges:

> *Those who live in the shelter of the Most High*
>   *will find rest in the shadow of the Almighty.*
> *This I declare about the Lord:*
> *He alone is my refuge, my place of safety;*
>   *he is my God, and I trust him.*
> *For he will rescue you from every trap*
>   *and protect you from deadly disease.*
> *He will cover you with his feathers.*
>   *He will shelter you with his wings.*
>   *His faithful promises are your armor and protection.*
> *Do not be afraid of the terrors of the night,*
>   *nor the arrow that flies in the day.*
> *Do not dread the disease that stalks in darkness,*
>   *nor the disaster that strikes at midday.*
> *Though a thousand fall at your side,*
>   *though ten thousand are dying around you,*
>   *these evils will not touch you.*
> *Just open your eyes,*
>   *and see how the wicked are punished.*
>
> *If you make the Lord your refuge,*
>   *if you make the Most High your shelter,*

no evil will conquer you;
  no plague will come near your home.
For he will order his angels
  to protect you wherever you go.
They will hold you up with their hands
  so you won't even hurt your foot on a stone.
You will trample upon lions and cobras;
  you will crush fierce lions and serpents under your feet!

The Lord says, "I will rescue those who love me.
  I will protect those who trust in my name.
When they call on me, I will answer;
  I will be with them in trouble.
I will rescue and honor them.
I will reward them with a long life
  and give them my salvation."

Connection Despite the Distance
June 26, 2003

I spent a heartfelt moment expressing my feelings in a four-page letter to my wife, emphasizing her immense importance in my life. Pouring my thoughts onto paper allowed me to articulate the depth of my emotions and the profound connection we share, even across the distance that separates us. After that, we shared an hour-long conversation over the phone, deepening our connection despite the physical distance.

Unexpected Twist
June 28, 2003

After completing a thorough cleanup of Kuwait during our police call (picking up trash), I'm about to sit down and write to my wife, sharing the happenings of the day. In a lighter moment, while at the soldier's tent, a surprising sight unfolded as I played football—a mouse unexpectedly leaped into the air, adding an unexpected twist to the day.

Memorial Service
June 29, 2003

Today, I'm reminded of the impending promotion board in August after attending an inspiring church service where the speaker presented a great message about being spiritually strong despite the physical limitations of the desert. Later that day, at Camp Wolf, a moving memorial service was held in honor of the men from a unit who unfortunately passed away. On a happier note, I received the last of my wife's care packages, which added a sense of comfort to the desert environment.

Grounded in Faith
July 1, 2003

My wife was delighted to receive the 4-page letter I sent, which took 11 days to arrive. Interestingly, my supervisor initiated a discussion intended at provoking me, as he had done with another NCO. In such moments, I've learned that choosing to distance oneself from negativity is often wiser than succumbing to negative thoughts and actions. In response, I had to draw on the wisdom of the Bible to navigate the situation. Staying grounded in faith and focusing on personal growth becomes paramount when faced with individuals aiming to cause harm or annoyance.

> ... choosing to distance oneself from negativity is often wiser than succumbing to negative thoughts and actions.

Rational Leadership
July 3, 2003

It's been observed that some soldiers are struggling to maintain personal hygiene, and as NCOs, we find ourselves unfairly blamed for this issue. It's crucial to recognize that emotions can run high in such situations, but it's essential not to react impulsively. Instead, as leaders, we need to stay rational, focusing on what's best for the individuals involved and the collective well-being.

On the other hand, the consistently high evening temperature of around 85°F presents a major challenge for all of us. Sleeping in such temperatures not only impacts our level of comfort but also adds another degree of difficulty to the already harsh conditions we face. The combination of scorching temperatures and the natural difficulties of our climate highlights the determination required to get through each day.

Contentment and Joy
July 4, 2003

Today brought a sense of contentment as I connected with my wife, who shared remarkable news of her dedicated efforts in losing 20 lbs. within just two months—a truly impressive achievement. Despite a bit of disappointment over the absence of fireworks during the 4th of July celebration, the excellent supper, particularly the ribs, was a notable highlight.

However, my thoughts go back to the anticipation of returning to the United States and reuniting with my wife, which adds a level of meaning and joy to these times of connectedness. This hopeful anticipation fuels my spirit and makes my heart full.

Sudden Changes
July 5, 2003

Today held promise, with the expectation of a shift change that would take me to Camp Doha, a visit I didn't mind at all. However, a sudden change to the roster, without prior notice, brought a sense of frustration. It's moments like these that remind me of the unpredictability of military life, where plans can swiftly change.

While it's easy to feel irritated, it's better to choose not to dwell on the negative thoughts, recognizing that some things are beyond our control. This mindset helps me maintain a positive outlook and prevents unnecessary frustration from dampening my mood.

Trust in God's Plans
July 6, 2003

The church service today was uplifting, emphasizing the importance of faith, especially during challenging times. It's a reminder that even in the face of difficulty, we shouldn't lose hope but we should strengthen our faith. It is important to focus on God's Word and trust in His plans for us.

In tough times or even during the good times, it's not just about persevering and moving forward with our lives, but also about having faith that God will never forsake us. This faith not only makes us stronger but also provides us with the physical, mental, and spiritual strength needed to navigate through uncertainties.

Shared Experiences
July 7, 2003

The morning began with a sense of fellowship at the soldier's tent, where I participated in a shooting video game on Xbox 360, finding moments of relaxation and connection with fellow soldiers. Following these shared experiences, I finished the morning by having a heartfelt conversation with my wife, closing the gap between us, even if just through words. These moments of connectedness and togetherness serve as crucial supports in the midst of the daily struggles we encounter.

Unforeseen Incidents
July 8, 2003

Today, I explored Camp Arifjan, a huge and vast base. However, our voyage was not without difficulties, as our Humvee developed a mechanical problem along the way. When attempting to move forward, the wheels suddenly backed up. Despite this incident, the trip provided an overview into the enormity of military infrastructure, as well as the unforeseen twists that can occur even during routine movements.

Withstanding Adversity
July 10, 2003

In an email, my wife reflected on the significant impact of faith in our lives, expressing gratitude for God's presence. She acknowledged the inherent challenges of enduring separation and highlighted the importance of the Holy Spirit in preserving our marriage during these trying times.

Her words served as a reassurance that, despite the hardships faced by many families during lengthy deployments, we were determined to withstand the adversities and save our love and memories that had shaped our relationship. We clung to those precious times in the face of a seemingly diminishing connection, determined not to let the strain of deployment compromise the relationship we had created.

Birthday in Kuwait
July 12, 2003

Today is my birthday, and I wanted to express my gratitude to the Lord for granting me another year of life and preserving my mental well-being. Although I'm spending the day in Kuwait, away from my wife, I'm taking the opportunity to relax, sleep, and enjoy some personal time. Despite the physical distance, I wish we could celebrate together.

I received a thoughtful gift from my church, Zoe Christian Center, sent on the 24th of May—it took 50 days to reach me. The fan included in the package will be especially valuable in this hot environment. Despite the distance, I consider it an outstanding celebration here in Kuwait, appreciating the blessings and the gestures of love from afar.

Tranquility
July 13, 2003

In today's church service, the sermon centered around the significance of having a vision. Stepping out of the tent afterward, I was greeted by intense heat, which made me feel like I was in a sauna.

MY FIRST DEPLOYMENT

The scorching conditions felt like walking past fire. Most of the day was spent in much-needed rest. Tonight, I am going to direct my creative energy into drawing a 1972 Chevy truck that was featured in a Chevy magazine, and I hope to find tranquility in this creative process.

Familiar Routine
July 15, 2003

After a comforting conversation with my wife this morning, we shared our excitement about the prospect of reuniting, though the separation does bring moments of sadness. In the midst of separation, our faith serves as a source of comfort, providing assurance that God will guide us through the challenges we face.

As I prepare for a visit to Camp Doha, I am embracing the familiar routine of my day in Kuwait. It's a moment to engage with the surroundings and navigate the daily tasks, finding a sense of normalcy in the midst of the unique challenges presented in this environment.

Moments of Reflection
July 17, 2003

This morning brought a refreshing change, with clean air and a gentle breeze. I took the time to delve into the pages of Our Daily Bread booklet, gaining insights that prompted reflection on various aspects of my life and faith. The opportunity to read this book has proven to be a source of gratitude, offering valuable lessons and moments of reflection.

PART I: THE FIRST SIX MONTHS

Recharging
July 20, 2003

I had a day filled with responsibilities, starting with Quick Reaction Force (QRF) duty, followed by serving as SOG for the motor pool in the evening. As a pleasant surprise, I received a DCU cap and flashlight. Once my shift concluded, I made sure to get plenty of sleep to recharge for the next day's tasks.

Anticipation
July 21, 2003

I started my day by having a wonderful conversation with my wife. After our conversation, I geared up for a workout at the gym tonight, aiming to stay physically active. On a brighter note, I received a postcard from my mom, sent out on July 14th, adding a touch of warmth to my day. However, the anticipation for the box sent on July 1st continues, and the delay only heightens the excitement for its eventual arrival.

Spiritual Connection
July 24, 2003

Receiving a care package from my family brought a mix of emotions. Due to a mix-up in the address, it took 24 days to reach me, as it initially went to another forward operating base (FOB) before being rerouted back to my location. Despite the delay, when I opened the package, its contents provided a comforting connection to home.

In a conversation with my mom, I learned that my brother, Timothy, gave his life to Christ, bringing me genuine happiness, and a sense of spiritual connection even in the midst of separation.

MY FIRST DEPLOYMENT

Trip to Doha
July 27, 2003

Today, I had the opportunity to return to Doha, and this time, I visited a place called the Marble Palace. The highlight of the day was observing people engage in volleyball, both in the water and on the sand, providing a refreshing break and a chance to unwind and connect in conversation. I discovered peace and tranquility during my one-day excursion, which provided an ideal break from my daily routine. The event provided enjoyable moments and a much-needed relaxation, allowing me to unwind and recharge.

Cherishing Every Moment
July 28, 2003

Having the opportunity to talk to my wife twice today made for a good morning. I prioritize these conversations as one can never predict disruptions like phone line disconnections or loss of internet signal, emphasizing the importance of cherishing every moment of connection despite the uncertainties.

Enlightening Meeting
July 29, 2003

The NCO meeting today provided an opportunity to discuss everyone's morale and well-being. It was an enlightening session in which the team's problems and positive qualities were expressed, fostering a sense of togetherness and understanding. Such sessions are critical in creating a welcoming atmosphere within the unit.

PART I: THE FIRST SIX MONTHS

Unique Adventure
July 31, 2003

This morning, I embarked on a unique adventure—I attempted to ride a real camel. Capturing the moment with photos, the experience was unforgettable as the camel knelt on all four knees, facilitating my climb onto its back. The intricate process involved the back legs rising first, followed by the front legs, creating a thrilling moment that felt like a precarious balance on the front of its neck. To commemorate the day, I purchased souvenirs for my mother and mother-in-law from the lively bazaar set up for everyone to explore and shop.

Me riding a camel for the first time. It was a thrilling experience.

Greater Things Await
August 1, 2003

Today was a meaningful day. I had an extended conversation with my wife, which brought her great joy. Later, at Joy Night in church, several ministers, including one who prayed with me, emphasized the importance of opening up my heart to let God work through me. Joy Night, a special church service event, aimed to uplift and share the Good News of Jesus Christ. The minister's words aligned with what I had been trusting God for. It assured me that I could open my mind to God's possibilities even during deployment. This strengthened my resolve, reassuring me that there were greater things awaiting me beyond the challenges of my current circumstances.

Shared Responsibility
August 2, 2003

During our conversation, my wife expressed a sense of overwhelming responsibility as she had to step into my shoes and manage tasks I typically handled as a husband, including ensuring the upkeep and maintenance of our vehicle. Despite the challenges, her commitment to maintaining our daily life and responsibilities brought a sense of unity and shared responsibility even while miles apart. I always keep her in my thoughts because it is important to maintain our connection and our commitment towards each other, and thinking about her instills my commitment to myself.

Resolving an Issue
August 3, 2003

This morning, I attended a church service that focused on the topic of false prophets, providing insightful discussions. A noteworthy observation from this week was the Kuwaiti construction method, where everything is meticulously crafted by hand without the use of power tools. Upon returning home after lunch, I found myself nominated to move into the soldier's tent to ensure it met the required standards. I identified and resolved an issue with the soldier's air conditioner, caused by a trash bag obstructing the vents. Witnessing their smiles and the improved functionality brought a sense of accomplishment, knowing I could enhance their comfort in the challenging environment.

Corporate Anointing
August 5, 2003

As I prepare for the promotion board, I find myself immersed in study materials, diligently reviewing the necessary content to excel. Later, I plan to delve into the insightful CDs that my wife thoughtfully sent me on "The Corporate Anointing." These resources not only aid my professional development but also serve as a reminder of the unwavering support from my loved ones back home. The opportunity to expand my knowledge and maintain a spiritual connection become a source of motivation during my deployment.

Trust and Faith
August 7, 2003

My garments clung to me like as if I'd been caught in a heavy shower today, despite the burning heat and humidity. As I reflected on the first year since my grandmother passed away, the gloomy weather appeared to echo the heaviness of what I was feeling.

Despite the difficulties, I find comfort in knowing that with God by my side, I shall endure and succeed. Trusting in God and believing in His plans helped me cope with every experience I had, the struggles, the negative comments about my faith. Everything.

MY FIRST DEPLOYMENT

Adapting in a New Environment
August 15, 2003

My NCOIC informed me about relocating to the Brigade Tactical Operations Center, confirming what the soldiers had mentioned a day earlier. This news brought a tinge of disappointment as I had established a sense of comfort in that area, with personalized setups that made it feel like home. Despite the change, I look forward to adapting and making the new environment equally welcoming.

Submitted Paperwork
August 25, 2003

Today has been quite manageable, work wasn't too demanding, but the temperature in the tent fluctuated, making it occasionally uncomfortable. I've submitted all the necessary paperwork to my NCOIC for the promotion process. Surprisingly, my body seemed to be adjusting to the heat, and I found myself sweating less. Additionally, a pleasant surprise awaited me as I encountered an NCO from my hometown, creating a connection that brings a touch of familiarity to this distant deployment.

Feedback and Expectations
August 27, 2003

During my time on Brigade duty, one of the NCOICs overseeing me efficiently conducted both my initial and monthly counseling sessions. It was a productive and organized process, providing valuable feedback and setting clear expectations for my responsibilities. These counseling sessions contribute to a structured and effective communication channel, enhancing the overall efficiency of our duties.

PART I: THE FIRST SIX MONTHS

Capturing Moments
August 28, 2003

Capturing moments with a digital camera this morning brought unexpected joy as I shared the pictures with my wife. Her enthusiastic reaction, a blend of excitement, surprise, and tears, prompted her coworker to rush over and share in the delightful moment. It's heartening to know that even miles apart, simple gestures can bring immense happiness and connection.

Labor Day
September 1, 2003

As I was ending a heartfelt conversation with my beloved wife, I learned she's battling a bad cold, and I sincerely hope for her speedy recovery. I assured her that I would keep her in my thoughts, sending warm wishes for a swift recovery, and promised to call her again soon to check on how she's feeling.

A letter from my sister, Lashannon, added a touch of warmth to my day. For dinner, it featured a delicious T-bone steak, leaving me with a content and satisfied stomach, ensuring a joyous evening ahead.

I also wrote a letter to my sister, Lashannon, sharing with her my experiences and also asking her about how they were all doing. It added a touch of warmth to my day.

Being Strong in the Lord
September 2, 2003

I wrote a letter to Lashannon today, updating her about the experiences I had these past few days and sharing about random things, expressing thoughts and maintaining ties that distance cannot break. These moments of spiritual reflection and communication with loved ones add a sense of purpose to my days here.

Drawing strength from the words of Philippians 4:13 (NLT), I reflect on the importance of being strong in the Lord during challenging times:

"For I can do everything through Christ,[a] who gives me strength."

Added Sweetness
September 5, 2003

Starting the day with prayers and Bible reading, I found solace in connecting with my faith. A conversation with my wife brought reassurance, and I took a moment to enhance the tent's comfort by placing bags of ice on the AC. Along with this thoughtful gesture, a delightful surprise arrived in the form of cookies and candy from my mom, coupled with a heartwarming letter, adding sweetness to my day.

Encouraging Connections
September 6, 2003

Returning from a trip in the Humvee to take a shower, I spoke with my wife, who is recovering from a cold, and I also received an encouraging email from my sister. These brief moments of connection add a sense of familiarity to my days in Kuwait.

The letters from home serve to reduce the distance between us and lighten the challenges of being away from each other. Each message and conversation are like a lifeline, reminding me of the love and support waiting for me on the other side of this deployment.

PART I: THE FIRST SIX MONTHS

Glimmer of Hope
September 7, 2003

In my discussion with my supervisor, a glimmer of hope emerged as he expressed his willingness to support my journey to the promotion board. Attending church proved to be spiritually uplifting, particularly as the speaker shared a powerful testimony about the transformative impact of God's saving grace on his life. This sense of encouragement and purpose in my daily life here in Kuwait is fostered by these positive developments.

Change in Weather
September 9, 2003

Feeling better with a day off on a Sunday and a change in weather, I just finished a reassuring conversation with my wife, who is also improving from her cold. Despite the distance, I wish I could be there to provide comfort and care for her. It's also noteworthy that the Bazaar made an incredible $60K profit yesterday. That's such a huge amount!

Preparing for Promotion Board
September 10, 2003

Continuing my preparation for the promotion board, my supervisor has been throwing various military questions my way. It's a challenging but essential part of getting ready for the board, and I appreciate the opportunity to reinforce my knowledge and readiness for the upcoming evaluation.

MY FIRST DEPLOYMENT

Winning Teamwork
September 11, 2003

We teamed up to tackle the ongoing issue of sand piling up in front of our tent. Our goal was to clear it away and fill sandbags with it. Our supervisor made it fun by timing us in a competition. I partnered with someone, and together, we filled nine sandbags in just two minutes which made us the winner. This not only strengthened our teamwork but also demonstrated our ability to adapt and overcome challenges.

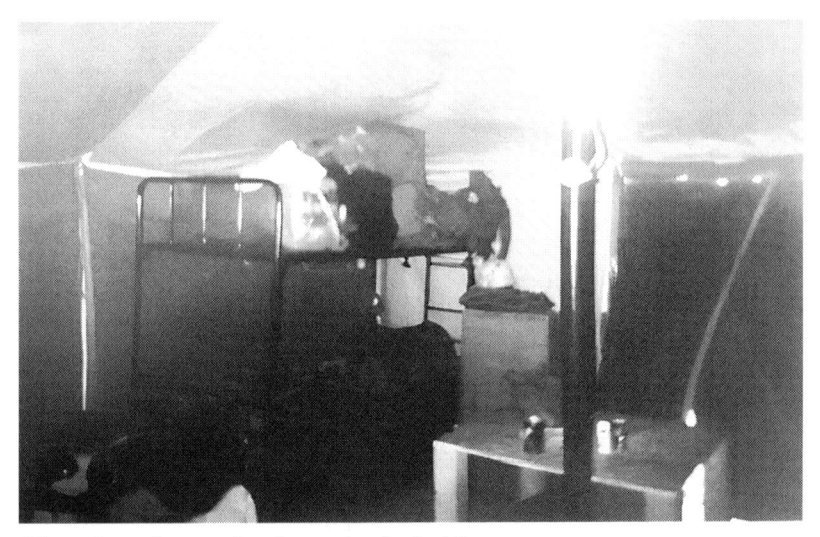

This what the inside of our tent looks like.

Study Questions
September 12, 2003

Today turned out well, and my supervisor handed me some study questions for the promotion board. I believed that looking into these questions would considerably improve my preparation, giving me the knowledge and confidence I needed to succeed. I'm

PART I: THE FIRST SIX MONTHS

always grateful for the guidance and support as I worked towards my goal.

Thoughtful Gesture
September 21, 2003

I attended church with one of the soldiers, and I hope he also found the experience as uplifting and spiritually enriching as I did. Sharing moments of faith can be a source of connection and encouragement in this challenging situation.

> Sharing moments of faith can be a source of connection and encouragement in this challenging situation.

Receiving a care package from my wife added a comforting touch to my day, carrying with it warmth and a sense of familiarity. I'm excited to check its contents, and I value the thoughtful gestures that bridge the distance and make this deployment more bearable.

Promotion Points
September 23, 2003

I headed to the internet café and worked on some correspondence courses. These courses are not only valuable for personal development but also contribute to earning promotion points. It was a productive way to utilize my time and work towards advancing my military career.

Routine Tasks
September 24, 2003

In the evening, my duties led me to Camp Wolf, where I assisted an NCO in catching a plane to Iraq. The responsibilities involved in ensuring a smooth departure added a sense of purpose to my day. After assisting the NCO, I headed back to camp and reflected on the significance of these routine tasks in contributing to the overall functionality of our operations here in Kuwait.

Road-Testing Humvees
September 25, 2003

Today, I was assigned to road-test Humvees at Camp Victory, assessing their functionality and performance. After a productive day, I had a conversation with my wife, and she reminded me to sample the food she sent before indulging in it. It's one of those moments that bring warmth and a touch of home to my experiences here in Kuwait.

Palace on the hill next to Camp Victory in Iraq.

PART I: THE FIRST SIX MONTHS

Eager to go Back Home
September 30, 2003

I started the day by running 2 miles to keep up with my fitness routine. Later, I received a heartwarming email from my wife expressing her anticipation for my return, which lifted my spirits. Afterwards, I shifted my focus to correspondence courses, which aimed to enhance my knowledge and skills while eagerly looking forward to being back home.

MY FIRST DEPLOYMENT

PRAYER

Dear Heavenly Father,

In the name of Jesus, we humbly gather before you, our hearts laid open, baring the depth of our emotions. Your divine guidance encourages us to seek you in moments of distress, finding solace in the reassurance of your Word:

"Turn to me and have mercy, for I am alone and in deep distress. My problems go from bad to worse. Oh, save me from them all! Feel my pain and see my trouble. Forgive all my sins." (Psalm 25:16-18, NLT)

We lift our voices in unity, reflecting on the poignant plea from Psalms. As we collectively face challenges and encounter adversaries, we resonate with the acknowledgment of numerous enemies and their vehement hostility:

"See how many enemies I have and how viciously they hate me! Protect me! Rescue my life from them! Do not let me be disgraced, for in you I take refuge." (Psalm 25:19-20, NLT)

We extend this prayer to those around us, our brothers and sisters, as they navigate their own tribulations. In their times of separation from loved ones, may the comforting embrace of your Word provide solace. Grant them strength to stay focused amidst challenges and adversity. May their faith remain unwavering, and may their gaze be steadfastly fixed on you.

In Jesus' Name, Amen.

PART II: THE SECOND HALF

This part explores the delicate balance between duty and personal connections, which will highlight my unwavering commitment to both my military responsibilities and the profound bond with my loved ones.

MY FIRST DEPLOYMENT

Starting the Day Right
October 1, 2003

I kicked off the day with a 2-mile run. After my 2-mile run, I proceeded to go to the dining facility for some food, followed by checking emails, and a reassuring call to my wife. Back in my room, I found solace in reading the Bible, grounding myself in faith and reflection.

Acknowledging Blessings
October 5, 2003

Today, I welcomed the day with gratitude, recognizing it as a precious gift and acknowledging the blessing of being alive and fully functional. Tomorrow marked the start of weight training, a step towards enhancing physical fitness and well-being, so I prepared myself by having the proper mindset and not focusing too much on anything stressful.

Unpredictable Nature
October 7, 2003

Selected for a task, I ventured to Camp Wolf to pick up a soldier bound for the Continental United States. The day extended into nighttime, which involved additional responsibilities, showcasing the dynamic and sometimes unpredictable nature of military duties.

Changes
October 9, 2003

An energizing upper body workout set the tone for the day. The departure of two NCOs back to the United States added a layer of change, emphasizing the unforeseeable nature of military personnel movements.

PART II: THE SECOND HALF

Postponed Trip
October 10, 2003

Conversations with my parents and wife helped to continue strengthening connections with my family. Due to logistical difficulties, the planned trip to the Kuwait mall was postponed, showing the adjustments required in military life. I checked my weight, and I weighed 160 lbs. and was pleased with my physical well-being.

Escort Duty
October 11, 2003

From 6:00 am to 6:00 pm, I was on escort duty, accompanying foreign nationals on their daily activities around the camp. A fellow rider from Egypt shared heartwarming stories, including a visible scar from a past encounter with a bullet in Iraq, bringing a sense of reality to those I encountered.

Significant Impact
October 12, 2003

In the morning, I attended church, and during the service, one of the speakers singled me out for a prayer. They emphasized the idea that an individual can make a significant impact, suggesting that God has plans to elevate me to a higher level in life.

Staying Vigilant
October 15, 2003

While returning from Camp Wolf, I encountered a challenging situation outside the gate as a car in my lane approached head-on. In response, I swiftly maneuvered to the far right lane to avoid any potential collision. The incident served as a reminder to stay vigilant on the road and prioritize safety.

Interesting Dream
October 17, 2003

After checking my email, I found a message from my sister sharing an interesting dream she had about me coming home early. Her words added a touch of positivity to my day, sparking a sense of anticipation and hope. Dreams can be a source of comfort and inspiration, especially when they align with our prayers and our thoughts.

Impactful Teaching
October 18, 2003

It was one of those days where the longing to be with my wife had already intensified. But I have accepted that some things are beyond my control, so I chose to surrender it to God and trust that He will work things out. I also attended church on Sunday and found the sermon to be impactful, focusing on Luke 14:26-33 (NLT) from the Bible. The teaching resonated deeply with me:

"If you want to be my disciple, you must, by comparison, hate everyone else—your father and mother, wife and children, brothers and sisters—yes, even your own life. Otherwise, you cannot be my disciple. And if you do not carry your own cross and follow me, you cannot be my disciple."

"But don't begin until you count the cost. For who would begin construction of a building without first calculating the cost to see if there is enough money to finish it? Otherwise, you might complete only the foundation before running out of money, and then everyone would laugh at you. They would say, 'There's the person who started that building and couldn't afford to finish it!"

"Or what king would go to war against another king without first sitting down with his counselors to discuss whether his army

PART II: THE SECOND HALF

of 10,000 could defeat the 20,000 soldiers marching against him? And if he can't, he will send a delegation to discuss terms of peace while the enemy is still far away. So you cannot become my disciple without giving up everything you own."

Collecting Supplies
October 21, 2003

In the morning, members of my unit arrived from Iraq to collect supplies, and it was a relief to see familiar faces. Additionally, I received a thoughtful care package from my wife, complete with a heartwarming card that brightened my day.

Brigade Duty
October 23, 2003

Today, I had to assume Brigade duty due to one of the NCOs being deployed to Iraq. It was an unexpected responsibility, but I adapted to the situation, ensuring the duties were covered in their absence.

Studies and Tasks
October 24, 2003

I started the day by completing two correspondence books and dedicated time to studying for the promotion board. As I navigated through my tasks, I held onto my faith and trusted that everything would fall into place.

MY FIRST DEPLOYMENT

Departing from Camp
October 26, 2003

My unit is departing from the camp to return to Iraq, and I anticipate following them in a few days. As this transition approaches, I find myself mentally and physically gearing up for the forthcoming journey, ensuring that all necessary arrangements are in place. The days leading up to our departure are filled with a blend of emotions, but there's a collective determination to face the challenges ahead.

Unwavering Support
October 28, 2003

Just finished a heartwarming conversation with my wife, and she's in good spirits, eagerly awaiting the end of my deployment. The anticipation of our reunion keeps our connection strong, and I'm grateful for her unwavering support and strength. Now, as the night settles in, I reflect on our talk and look forward to the day we can share each moment in person.

Chilly Weather
October 31, 2003

It's Halloween today, and I found myself on tower guard duty at the front gate as the NCOIC. The shift is a manageable six hours, but the day and night bring a relentless sandy wind, creating a turbulent, yet chilly atmosphere. Despite the eerie weather, there's a sense of duty and responsibility that keeps me focused on the task at hand.

PART II: THE SECOND HALF

Cozy Home
November 2, 2003

Today unfolded as a good day with multiple conversations with my wife, bringing a sense of affection and connection across the distance. Her description of the computer room at our home, which was a cozy space with curtains at the window, painted a vivid picture in my mind. As the day wound down, I turned to the solace of reading the Bible before heading to bed, finding comfort in the words that anchor my faith.

Source of Strength
November 4, 2003

The joy of knowing that my wife received my letter yesterday fills my heart with joy. Her love is a constant source of strength, and I cherish the thought of returning to her arms. In the midst of the challenges, her presence in my life is a beacon of hope and strength.

Sudden Change in Plans
November 7, 2003

After completing a shift on gate guard duty at 7:00 am, I'm asked to pull duty at Brigade TOC at 8:00 am. The sudden change in plans is a reminder of the unforeseen nature of military life. However, the OIC reassures me not to worry about duty, providing a brief respite.

MY FIRST DEPLOYMENT

Fierce Sandstorms
November 10, 2003

When I talked to my wife, she informed me that she returned safely to Oklahoma from Baltimore, which made me feel relieved. However, in the motor pool, a fierce sandstorm blanketed everything, leaving sand in every nook and cranny of the area. Despite the discomfort, the strength of our spirits remained unwavering.

Veteran's Day
November 11, 2003

It's Veteran's Day, and the weather was erratic with heavy rain, but shortly thereafter, the sun appeared bringing both warmth and humidity as well. As I reflected on the celebration of Veteran's Day, I realized the importance of camaraderie among veterans and the support given by the people around them.

New Assignment
November 13, 2003

News arrived that I would be heading to Camp Arifjan to escort a vehicle, introducing a change in my routine. The adaptability required in military life becomes apparent as I prepare for this new assignment, demonstrating the importance of adaptability and readiness in navigating the ever-changing deployment situation.

Suffering from a Cold
November 15, 2003

Today I felt under the weather as I was suffering from a cold, I bought medicine to help with the symptoms I was experiencing. Despite the physical discomfort, I was determined to overcome any difficulty, and I looked forward to better days ahead.

PART II: THE SECOND HALF

> Despite the physical discomfort, I was determined to overcome any difficulty, and I looked forward to better days ahead.

God's Intervention
November 16, 2003

Despite my weakened condition, I had to go to the motor pool this morning. One of my bosses, to my surprise, used vulgar language, causing me to think about a measured response. Before I could address the situation, he surprisingly apologized. It felt like God understood my situation and intervened in the situation on my behalf.

Rare Treat
November 18, 2003

In the morning heat, I had a heartfelt conversation with my wife, finding solace in the consistency of our communication. The day unfolded at a slower pace, culminating in a dinner that brought joy through a flavorful T-bone steak, a rare treat, in the somewhat monotonous routine during my deployment.

Captivating View
November 19, 2003

As I visited Camp Doha, the beauty of Kuwait City across the ocean captivated me. The landscape, with its huge expanse of sand blending into the buildings, created a breathtaking view that was both unfamiliar and fascinating.

Meeting Acquaintances
November 20, 2003

I packed my bags for Camp Arifjan, and I was surprised to see my first supervisor from my initial duty station, now retired. I then settled into a warehouse filled with bunk beds that marked a shift in my living conditions, reflecting the necessity of being adaptable in military life.

Stable Routine
November 23, 2003

I returned from breakfast, a stable routine in the day, which provided me a sense of control and peaceful solitude in my life. The morning meal became a cherished part of my daily life, offering a moment of calm before the bustling activities of the day.

Vital for Survival
November 24, 2003

Going to the gym and having regular calls to my wife played vital roles in my survival during deployment. Engaging in exercise not only helped alleviate stress but also served as a consistent outlet for maintaining my overall physical well-being. These practices became essential routines during my deployment.

PART II: THE SECOND HALF

Thanksgiving Day
November 27, 2003

It's Thanksgiving Day today, my wife received a letter from me, and although she felt a bit down about my absence, I encouraged her to stay strong and keep the faith. While attempting to pray for both of us, the phone connection became unstable. Despite the challenges, the Thanksgiving lunch was delightful, featuring ham, turkey, candied yams, green beans, Cornish hen, and two cups of sweet tea.

Joy and Thoughtfulness
November 28, 2003

A hearty conversation with my wife made up for an awesome Thanksgiving, even though we were apart. The thoughtful card and flowers she sent me served as gentle reminders of our connection and love, bridging the gap created by distance.

Dashika's Birthday
November 29, 2003

It's my wife's birthday today, and the longing to be home to celebrate with her was evident. Though we were not together, we found solace in understanding the sacrifices made for our shared journey, making the anticipation of reunion even more meaningful.

Erratic Weather
December 1, 2003

Doing laundry was top in my list of chores today, second was to talk to my first sergeant. He shared that he was going back to Iraq this morning. The weather was erratic today and it has been raining almost all day. It was coming down so extreme that it was hard to get to the showers with all the puddles on the ground.

Near-Death Experience
December 3, 2003

Getting soaked on my way back from the shower, I navigated through heavy rain, highlighting the struggles of staying dry. Later, another Humvee came within inches of hitting my Humvee on the passenger side. I must say it was such a tense and nerve-wracking experience.

Christmas Parade
December 10, 2003

The small Christmas parade brought a festive atmosphere, lifting everyone's spirits. Following the parade, I had the pleasure of talking to my wife, complimenting her as the most beautiful woman in the world, which added excitement to her day.

Undeniable Anticipation
December 12, 2003

In an engaging conversation with my wife, I learned that she's doing well, but there's an undeniable anticipation for the moment when I return home. As we shared our thoughts and feelings, the prospect of being together again became a source of mutual excitement and comfort.

Contrast in Beliefs
December 13, 2003

Even though my supervisor was reluctant to embrace change, I reiterated my commitment to pray for him and hoped for a positive transformation in his life. Our conversation revealed a contrast in beliefs, but I held onto the belief that the power of prayer could bring about meaningful shifts in the lives of those open to it.

PART II: THE SECOND HALF

Risks and Uncertainties
December 16, 2003

    The interruption during our flight to Balad, Iraq served as a stark reminder of the persistent risks and uncertainties that loomed in the region. The unexpected threat forced us to reroute and delay our plans, highlighting the unstable nature of the environment we navigated daily. Once the danger had passed, we could finally land, highlighting the ever-present challenges of operating in such a high-stakes setting.

The view in the Balad Air Base in Iraq.

Christmas Day
December 25, 2003

    Celebrating Christmas without my wife is bittersweet, but I find solace in the blessing of seeing another holiday season. Throughout the day, my responsibilities included shuttling soldiers to and from tower guard duty or back to their living areas. While the absence of loved ones is felt, I remain focused on the future, eagerly anticipating the day when I can celebrate these moments with my wife and family once I return home.

Status Red
December 29, 2003

Today we went out on a convoy with a "Red" weapon status, which means one round left in the chamber, and on safe mode. We finally got to our very small area, with no running water and chow hall. The highway we went down was called Main Supply Route Tampa. There was trash all over the roadway, blown-up tanks, and rockets on the ground.

End of a Busy Chapter
December 31, 2003

As the year came to a close, I find myself reflecting on the challenges and blessings it brought. Dressed in my DCU, I spent New Year's Eve watching NCAA Football with my tent mates, marking the end of a busy and challenging chapter of my deployment experience.

Welcoming the New Year
January 1, 2004

Grateful for another year, I thanked the Lord for the opportunity to see it unfold. Despite missing my wife dearly, I welcomed the new year with a sense of hope. Taking a moment for personal care, I washed my body, appreciating the simple joys that bring a touch of normalcy to my days.

PART II: THE SECOND HALF

Busy Night
January 3, 2004

A challenging night of heavy rain resulted in leaks in the tent. Resourcefully using trash bags to adjust the roof, I managed the setback. Later, heading to Camp Cedar II in Kuwait to assist with vehicle transfers, I reflected on the diverse experiences that deployment entails.

Harsh Reality
January 4, 2004

Arriving at Camp Cedar II, the heartbreaking scene of Iraqi children pleading for food stirred deep emotions within me. Witnessing their struggle highlighted the harsh realities faced by the local population and added a layer of compassion and empathy to my experience in this foreign land. Enduring a cold night, I found comfort in a warm shower and the chance to email my wife afterwards.

A group of kids waving at us.

Stories and Updates
January 6, 2004

Today's highlight was a conversation with my parents. Learning about my dad's successful deer hunt and the updates on my brother's singing provided a welcome glimpse into the familiar, offering a sense of home despite the distance.

Flooding and Persistent Rain
January 8, 2004

Facing persistent rain, I prepared to return to Camp Cedar II in Iraq. Flooded relay points added a layer of complexity to the journey, showcasing the unpredictable nature of weather conditions in the region. I felt nervous, but I chose to lift up all my worries and concerns to the Lord.

Risks and Reminders
January 9, 2004

Leaving Camp Virginia, I spoke to my wife before the journey back to Camp Cedar II. A brutal accident along the way served as a stark reminder of the risks associated with travel in a conflict zone, offering a sobering perspective on the fragility of life.

Ongoing Adjustments
January 10, 2004

Beginning the day with prayers and Bible scriptures, I faced the challenges of a muddy Camp Cedar II. The absence of a phone center, with only internet access, added another layer to the ongoing adjustments required during deployment.

PART II: THE SECOND HALF

Visual Reminder
January 13, 2004

Having a conversation with my wife today brought mutual joy, and it struck me when she was surprised that I've been away for nearly a year. Additionally, witnessing a Heavy Expanded Mobility Tactical Truck (HEMTT) that we had worked on, now restored, served as a visual reminder of the challenges and transformations that occur in the course of our deployment.

Easy Day
January 14, 2004

Having a relatively easy day, I observed non-functional Iraqi jets, a tangible reminder of the region's history. After a casual game of NCAA football, losing by a touchdown initially, I redeemed myself in the second round with a decisive victory, concluding the day on a positive note.

Familiar Faces
January 16, 2004

My day involved enhancing the protection of one of the HEMTT vehicles by installing steel covers to guard against IED and small firearms. As a surprising and heartwarming encounter, I ran into a friend from the church I attended in Germany, who now worked in the dining facility, adding a touch of familiarity to the day.

Training Accident
January 17, 2004

I started the day with reading the Daily Bread booklet and the Bible. However, I heard the news of the passing of a valuable NCO due to a training accident. Just a few days ago, he had expressed his commitment to assist me in preparing for the promotion board, showcasing his selfless dedication to supporting others beyond his immediate responsibilities. His absence was deeply felt, and he will be remembered with respect and gratitude.

Crisis Briefing
January 18, 2004

I talked to my family back in South Carolina and they're doing fine. Also today, we all had to go to a crisis briefing about the loss of one of our comrades. As we shared our memories and reflections during the briefing, a strong sense of camaraderie emerged, reinforcing our commitment to lean on each other for support throughout these challenging times.

Traveling to Camp Cedar II
January 20, 2004

We prepared to go to Camp Cedar II tomorrow, which will take us through the outskirts of Baghdad, Iraq. The anticipation of the road trip brought a mix of excitement and vigilance, as we will navigate through the diverse landscapes of this region. The day after, we traveled to Camp Cedar II and made it safely down through the Sunni Triangle without an incident.

Multiple Accidents
January 22, 2004

While our supervisor handled some business, we efficiently downloaded the HEMTT. During the process, I had a chance encounter with a fellow soldier from my hometown, bringing back

PART II: THE SECOND HALF

memories of our first duty station together. Moreover, yesterday, I saw this tractor-trailer that ran into the back of another trailer and almost cut the cab in half. Then we had a soldier from another section flip a Humvee from driving too fast. This made me realize that during deployment, we should always be careful and be prepared, since accidents could happen any time.

A Fuel Tanker stopping in front of an accident between two tractor-trailers.

"

> This made me realize that during deployment, we should always be careful and be prepared, since accidents could happen any time.

MY FIRST DEPLOYMENT

Continuous Rainfall
January 23, 2004

My morning started early at 4:00 am, and the day brought continuous rain until around 8:00 am. The amount of rainfall was unprecedented, exceeding what they typically receive in an entire year. The prospect of tomorrow's journey to Balad Air Base loomed ahead, adding a layer of anticipation to the already weather-affected atmosphere.

Stark Reality
January 24, 2004

Upon our safe arrival at Anaconda, the weather was refreshingly cool. However, the stark reality of poverty became apparent as we passed houses constructed from discarded trash found in the dump. Disturbingly, some local children expressed their disgust of us by making offensive gestures as we drove by, serving as a grim reminder of the challenges and tensions in the region.

A picture of me in Anaconda, Iraq.

PART II: THE SECOND HALF

Beginning of Preparations
January 27, 2004

Among the drivers, a senior officer shared that March would mark the timeframe for our departure from Iraq, signaling the beginning of preparations for our return to our duty station. The news injected a sense of anticipation and planning into our ongoing operations.

Hazardous Situation
January 29, 2004

In a gesture of love and connection, I sent my wife a Valentine's card and an email straight from the heart. Safely reaching our relay point, our supervisor's preference for constant movement was evident as he chose to ride over idle moments. However, a potentially hazardous situation unfolded as I disconnected a trailer from a 5-ton truck, and the trailer attempted to roll forward, posing a risk of pinning me between the truck and the trailer.

Guard Duty
January 30, 2004

Despite stepping into a mud-filled hole before tower guard duty, the day unfolded nicely. I finally could open the care package my wife sent, revealing a pile of goodies and two beautiful cards that warmed my heart. Reflecting on her support, I couldn't help but feel immense gratitude. Adding a surreal twist to the day, the MP set a tractor-trailer on fire, prompting interest from local Iraqis who sought to salvage parts from the burning vehicle.

MY FIRST DEPLOYMENT

Fragile Balance
February 1, 2004

A conversation with my wife while I was at Camp Scania provided a moment of relief and connection. The Defense Service Network phones proved reliable for communication with loved ones, but the caveat of potential shutdown during sandstorms or critical events was a constant concern, emphasizing the fragile balance between maintaining connection and ensuring operational security. The careful management of these networks highlighted the complex issues of deployment, particularly regarding communication.

Varied Responsibilities
February 2, 2004

After a shift on tower guard duty, a simple dinner of barbecued chicken added a touch of comfort. Engaging in intellectual pursuits, I delved into a book about George Washington Carver while simultaneously studying the Fort Sill guide. Balancing duties, from tower guard to personal development, painted a picture of the varied experiences during deployment, where moments of quiet reading and studying coexisted with the responsibilities of military life.

Struggles and Demands
February 3, 2004

Morning nourishment in the form of Ritz crackers with peanut butter preceded a day of physical training, highlighting the resilience required for the strenuous demands of deployment. The struggles combined with regular physical exercises like pushups among fellow soldiers exact a toll on physical fitness. The day's routine, fusing physical exertion and intellectual pursuits, offers a glimpse into the blended discipline and personal growth experienced during deployment.

PART II: THE SECOND HALF

Disappointing Decision
February 5, 2004

A refreshing night's sleep set a positive tone for the day, marked by tower guard duty and a subsequent weightlifting session. The casual mention of a supervisor's decision not to send anyone to the board in March added an unexpected layer of disappointment, emphasizing the uncertainty and fluctuating expectations that often characterize military life.

Emotional Highs
February 6, 2004

A trip to Camp Scania brought the joy of hearing my wife's voice, a cherished connection in an environment devoid of running water, phones, or internet. The recovery of my lost Gerber by a battle buddy added a touch of serendipity, serving as a small yet meaningful event in the midst of deployment challenges. The day summed up both the emotional highs of communication with loved ones and the unexpected moments of camaraderie.

Intricate Balance
February 10, 2004

Amid the routine of daily activities, including a shower, video games, and physical exercises, the deployment experience revealed its unique dynamics. Video games became a recreational outlet, showcasing the adaptation to new routines and the significance of finding moments of joy and distraction. The scene touched on the blend of military duties and personal pursuits combined.

MY FIRST DEPLOYMENT

Changing Landscape
February 11, 2004

A weather-centric morning on the 2:00 am to 4:00 am shift marked the challenges posed by dust storms, emphasizing the unpredictable nature of the environment. A visit to Camp Scania, though brief, allowed for a precious conversation with my wife. The impact of new leadership dynamics and their effect on communication illustrated the ever-changing landscape of deployment experiences.

Military Hierarchy
February 12, 2004

Post-shower rejuvenation set a positive tone for the day, including a tower guard shift in warm weather. The involvement in operational schemes, such as determining the turning radius of military vehicles, highlighted the multifaceted nature of responsibilities. An episode with a lieutenant, including an eventual apology, provided insights into the intricate dynamics within the military hierarchy.

Moment of Introspection
February 13, 2004

A day marked by quiet reflection, accompanied by listening to CDs from home, offered a respite from the usual hustle. Despite deployment difficulties, the sermon's themes regarding life preparation struck a chord with me, as it offered a chance for reflection and spiritual connection. The way the calm day was contrasted with the military lifestyle captured the complex feelings that were felt at this time.

PART II: THE SECOND HALF

> Despite deployment difficulties, the sermon's themes regarding life preparation struck a chord with me, as it offered a chance for reflection and spiritual connection.

Valentine's Day
February 14, 2004

Valentine's Day at Camp Scania allowed for a heartfelt call to my wife, a poignant acknowledgment of the profound connection amid physical separation. The evening's activities, involving washing and a game of Dominos, showcased the simplicity and camaraderie sought in the midst of deployment, creating a semblance of normalcy and connection during special occasions.

Free Time
February 15, 2004

This morning is very cold and windy. Compared to the weather the day before, it's much better. I had free time today, so I listened to a fourth CD from my wife, which talked about "Corporate Anointing." Since I did not have that many things to do, I chose to relax and do whatever I could today. It's a good thing we have time to do what we want every now and then.

MY FIRST DEPLOYMENT

Praying for Soldiers
February 16, 2004

I just got back from Camp Scania. I immediately called my wife, and fortunately, I was able to talk to her for 30 minutes, and then to my parents and my brother. He said that he was praying for me and the other soldiers. I felt touched when he included me and the other soldiers in his prayers. Everyone here needs all the prayer we can get. After talking to my family, I went back to our RP, since I had to serve chow to everyone. It was not a tiring activity, but it is part of the usual tasks I have.

Anticipation of Good News
February 17, 2004

The news of my supervisor holding my promotion packet this morning filled me with anticipation and excitement. The prospect of attending the promotion board and returning to the United States with an elevated rank added a sense of achievement and motivation, marking a positive milestone amid the challenges of deployment.

Proactive Step
February 18, 2004

My supervisor took the proactive step to start my Noncommissioned Officer Evaluation Report (NCOER) in order to record my performance while on deployment. It gives us something to look forward to when the first sergeant mentions possible replacements arriving around the middle of March, though hopefully, the schedule stays the same. The visit from our sergeant major and the new sergeant major from the National Guard also added a layer of inspection and surveillance on the site.

PART II: THE SECOND HALF

Extensive Measures
February 21, 2004

Observing an engineering unit detonate 5 antipersonnel mines during my tower guard duty provided a firsthand glimpse of the ongoing efforts to ensure safety in our vicinity. The distant flash, approximately 2.5 miles from our RP, highlighted the extensive measures taken to secure the area and neutralize potential threats.

Transitioning to a New Responsibility
February 25, 2004

As I began the process of sending my final letter to my wife and parents before transitioning the responsibility to the incoming unit, I learned about a Humvee accident involving one of our section chiefs. Seeing pictures of the aftermath, I couldn't help but feel grateful for the miracle of his survival. I need to reinforce a safety mindset at all times.

Not in the Duty Roster
February 26, 2004

Surprisingly, my name was absent from the duty roster this time, and it caught me off guard since I'm usually assigned various tasks. A duty roster is essentially a schedule that allocates specific responsibilities to individuals within a group or organization. Taking advantage of this rare break, I found solace in the opportunity to relax and enjoy some personal time.

MY FIRST DEPLOYMENT

The Perfect Storm
February 29, 2004

Earlier today started well, but it took a somber turn when a truck driver working on his vehicle got hit, reportedly flying about 40 feet and succumbing to his injuries. In response, some of our team members were tasked with taking one of our vehicles to recover the body and bring it back to our area. The medical personnel were contacted, and they arrived at our location to pick up the body for further procedures.

Despite the events, I found a moment to capture a photo of an Apache helicopter flying by the tower during my duty. Additionally, I accomplished finishing the book, "The Perfect Storm" amidst the events of the day.

Harsh Conditions of Deployment
March 3, 2004

The day brought hotter temperatures, worsening the flea and gnat situation. I now have itchy bumps on my arms from yesterday's bites. Despite feeling uncomfortable, I covered exposed skin and applied anti-itch cream with aloe for relief. Dealing with challenges became a daily routine as I navigated through the harsh conditions of deployment.

Preparations for My Return
March 5, 2004

In a talk with my wife, I shared the exciting news that I would be home in 5 weeks. Overjoyed, she began preparations for my return, discussing plans and places to visit with family. The anticipation of reuniting added a positive note to our conversation, making the distance feel more manageable.

PART II: THE SECOND HALF

Becoming an Uncle
March 8, 2004

During a visit to Camp Scania, I spoke with my parents and learned that my sister is 10 weeks pregnant with their first child. The prospect of becoming an uncle brought a sense of joy, and we discussed the anticipation of welcoming a new addition to the family.

Clean Uniform
March 10, 2004

Preparing for my departure from Iraq, I took the time to wash all of my dirty TA-50 and ensured I had a clean uniform and a fresh haircut. The attention to presentation reflected my eagerness to return home and be reunited with my wife.

Comforting Conversation
March 12, 2004

A conversation with my wife revealed that she wasn't feeling well and sounded weak. I prayed for her, and the next day, she expressed feeling better. Our talks bridge the distance and provide support during challenging times.

Improving Health
March 13, 2004

Continuing to call my wife regularly, I found solace in the ability to connect with her despite the physical separation. She mentioned the day before that she was not feeling well. Then she shared that she was now feeling better, so it brought me relief. I then thought about being deployed in a different country is not an easy thing, so the consistency of communication became a crucial lifeline during my deployment.

MY FIRST DEPLOYMENT

> ... the consistency of communication became a crucial lifeline during my deployment.

Quiet Sunday
March 14, 2004

A quiet Sunday ensued as the unit replacing us inspected the area. Recognizing familiar faces from my last unit, I exchanged greetings and inquired about their experiences since our last deployment together. This inspection made me remember that it's almost time for me to go home. It made me excited and glad that I would be able to see my wife again.

Need for Caution
March 15, 2004

As some soldiers went to Camp Scania for financial matters, I organized my bags in anticipation of leaving. While descending from tower guard duty, a soldier's weapon caught the handrail, causing my plate of chicken to fall. Fortunately, I avoided a potential fall, emphasizing the need for caution in the challenging environment.

Area Inspection
March 16, 2004

A National Guard unit from South Carolina inspected the area they would soon take over. The visit marked the beginning of the transition, emphasizing the cyclical nature of deployments and the constant flow of units in and out of duty stations.

PART II: THE SECOND HALF

Surreal Experience
March 17, 2004

Today marked a special moment as I saw my wife for the first time on the internet. Her beauty overwhelmed me with excitement, making the reality of coming home even more surreal. The sight of her face after almost a year of being apart brightened my day and became a cherished memory fueling my anticipation for the upcoming reunion.

Smooth Transition
March 18, 2004

The unit from South Carolina has arrived, and we are providing guidance on daily operations. By 8:00 am tomorrow, we'll no longer be in charge. My supervisor, with his wealth of experience, is actively assisting them with valuable insights and suggestions, ensuring a smooth transition in the handover process.

Christmas Card
March 21, 2004

We departed for Camp Cedar, and we endured a two-hour journey to a tent without heating, only AC. The following day, the motor pool activity was somewhat disorganized, but I received my Christmas card from my mom, sent in December, providing a comforting connection despite the challenging weather.

Relentless Bugs
March 24, 2004

Battling relentless bugs, their bites have left me itching and bleeding. A fellow soldier remarked on my humility, and I shipped

MY FIRST DEPLOYMENT

my 61-pound foot locker back to the States, costing $37.60. The discomfort contrasts with the excitement of getting home.

Memorial Sign Dedication
March 25, 2004

Exploring the ancient Ur temple in Iraq, over 5,000 years old, was a unique experience. We also participated in a memorial sign dedication for a fallen comrade. Departing for Camp Arifjan, Kuwait, to wash our vehicles for shipment home, the journey was thankfully smooth, allowing us to rest inside our vehicles in the sterile area upon arrival.

The Ur Temple where Abraham started his journey.

PART II: THE SECOND HALF

Item Checks
March 28, 2004

Concluding sensitive item checks and turning in our belongings, we woke up feeling congested, likely due to black mold in the tent and changing weather. Visiting the Bazaar provided a welcome distraction, and a small cellular phone's unexpected shock when I touched it added a humorous touch to the day.

Visiting a Classmate
March 31, 2004

Taking a moment from the busy schedule, I visited a high school classmate who was still in the country. We reminisced about shared memories, reflecting on the paths we've taken since high school. Seeing a familiar face in such a distant environment brought a sense of connection and nostalgia.

Redeployment Briefing
April 2, 2004

Attending a redeployment briefing, I retrieved my freshly cleaned DCU, preparing for my return home. Learning to use a Palm PC and completing a medical card added a practical touch to the day's activities.

Awards Ceremony
April 3, 2004

The award ceremony recognized platoon sergeants and above with Bronze Stars, while others received an Army Commendation Medal, and a Battalion Coin. The chaplain emphasized the temporary nature of the money earned, serving as a reminder of the sacrifices made throughout the deployment in a moving way.

PRAYER

Dear Heavenly Father,

We come before you, offering our honor and glory. We take joy in the truth from Romans 5:3-5, NLT:

"We can rejoice in problems and trials, for they develop endurance, strength of character, and a confident hope of salvation. This hope will not disappoint because of the deep love you have given us through the Holy Spirit."

We extend this prayer to all seeking solace in your Word. In their challenges, may they find the ability to rejoice amidst trials, knowing that endurance is cultivated through adversity. May the process of developing strength of character empower each individual. Let the confident hope of salvation be a guiding light, assuring them that it won't lead to disappointment.

Lord God, we collectively honor you for the immeasurable love you shower upon our lives. In gaining strength, we take pleasure in your unwavering support. You are our shield and our peace. May this realization resonate deeply within reflecting hearts.

In Jesus' Name, Amen.

PART III: GOING HOME

This part brings my year-long deployment to an end, capturing his intense excitement for my impending reunion with my wife. Readers may relate to my reflections on the obstacles, struggles, and unwavering commitment that characterized my service in this section.

MY FIRST DEPLOYMENT

Good News
April 4, 2004

Around noon today, I received the news that I'm on Chalk #1, the first to depart and head back to the United States. As the thought of coming home became a reality, the revelation sparked a flood of emotions and preparations for the journey ahead. The dreams I had of seeing my wife in person and being able to hug her and talk to her physically was now within reach.

Imminent Reunion
April 5, 2004

Today, in a heartwarming conversation with my wife, the shared excitement about our imminent reunion filled the air. The excitement of seeing each other in only a few days had filled our words with enthusiasm, turning the countdown into a tête-à-tête full of love and longing. As our talk continued, the simple exchange of anticipation became a blanket of comfort and peace. This was what I have been waiting for, and I am finally going home!

Preparing to Go Home
April 6, 2004

This marks the day the Lord has blessed me to embark on the journey of reuniting with my beautiful wife. As I prepare to go, the anticipation and thankfulness in my heart are real, and I take with me the assurance of heavenly blessings that will pave the way for our happy reunion.

PART III: GOING HOME

A photo of us as we wait to be released so we can see our loved ones.

Back Home
April 7, 2004

I finally arrived back at my duty station, where my million-dollar wife was waiting. When I first saw her, all the struggles and difficultiesfaded away, and in that poignant moment, the chapters of separation were replaced by the unwritten pages of togetherness. The priceless love we shared, built on resilience and patience, became the anchor that held us steady through the unpredictable tides of my deployment. But the Lord was the one in control. He protected

Seeing my wife for the first time after my deployment. The joy in our faces is shown as we held each other once again.

103

MY FIRST DEPLOYMENT

me and stayed with me during my deployment. Without His presence, I would have succumbed to my anxious thoughts and my feeling of loneliness.

> But the Lord was the one in control. He protected me and stayed with me during my deployment.

PRAYER

I lift my heart to You, embracing a trust that goes beyond what my eyes can perceive. In this journey of faith, I believe that You will provide the comfort I seek. My trust is anchored in You alone, not in the uncertainties of humanity.

As Your Word guides me, I commit to sharpening my ways with its wisdom. I choose obedience even in moments of confusion, trusting that You hold the master plan. Your Word reminds me:

"Trust in the Lord with all your heart; do not depend on your own understanding. Seek his will in all you do, and he will show you which path to take." (Proverbs 3:5-6, NLT)

I commit to seeking Your way, relinquishing dependence on my own desires and emotions. Lord, I surrender my feelings and wishes to You, trusting that You will fulfill every need. Guide my steps, and extend Your protection over me, my comrades, and my family, shielding us from harm.

In Jesus' Name, Amen.

CONCLUSION

In this recounting of my deployment journey, I have reflected and thought about many things during my whole deployment—the struggles, the excitement, the things I learned, and all the experiences I had are written in this book. The desire to be with my wife accompanied me through the scorching days and dusty storms. Despite this, the help and support I received became a pillar of strength for me. I found strength in their presence, whether it was the camaraderie of those by my side during deployment or the everlasting support and encouragement from family and friends back home.

Each adversity becomes a part of my experience of working through the struggles with perseverance. Yet it was my wife's letters and phone conversations, the voices of my family, and the unwavering encouragement of friends and colleagues that fueled my hope and strength. Their help transformed the everyday into accomplished moments which created a sense of unity that surpassed physical distance.

As I reflect on my first deployment, I carry with me not only memories of hardship but also a profound gratitude for the network of people who became my anchors. Their unwavering belief and encouragement turned the pages of challenge into a narrative of strength and resilience.

When I think back on my first deployment, I remember not only the hardships, but also the strength and perseverance I gained from all the experiences I had, but I also remember the people who became my anchors. Their unshakable faith and support transformed the feeling of adversity into strength and resilience. Though that yearning lingers, their presence, in many ways, guaranteed that I emerged from my first deployment unscathed. The Lord's presence in my whole deployment, as well as my connections and the love that surrounded me, kept me strong and made me feel protected.

I hope this book, which also doubles as a recounting of my days of deployment, has inspired you or made you feel loved and protected by God. May it serve as a reminder that struggles will always be present, but with the Lord, your support system, family, and friends by your side, you will persevere and thrive. Always make the Lord a part of your life, and His presence will be a comforting companion every single day.

Prayers and blessings for everyone.

- **Ronald Bethune**

A portion of the proceeds generated from this book benefits Engage Your Power, Inc.

ENGAGE YOUR POWER, INC. PROVIDES HOME IMPROVEMENT SERVICES TO VETERANS AND ASSOCIATES WHO ARE LESS FORTUNATE TO PROVIDE FOR THEMSELVES. OUR SERVICES INCLUDE PROGRAM DEVELOPMENT THAT WILL BE FUNDED BY GRANTS AND DONATIONS TO IMPROVE THE LIVING SPACE AND REDUCE THE STRENUOUS HEADACHE AND COST OF HOME REPAIRS THAT MAY SEEM ENDLESS.

engageyourpowerinc@gmail.com

580-678-5226

Made in the USA
Columbia, SC
15 November 2024